The ESSENTIALS® of

REGISTERED TRADEMARK

CALCULUS III

Staff of Research and Education Association,
Dr. M. Fogiel, Director

This book covers the usual course outline for Calculus III. Earlier topics are covered in *"THE ESSENTIALS OF CALCULUS I"* and *"THE ESSENTIALS OF CALCULUS II"*.

Research and Education Association
61 Ethel Road West
Piscataway, New Jersey 08854

THE ESSENTIALS OF CALCULUS III ®

Printed in the United States of America

Library of Congress Catalog Card Number 87-61815

International Standard Book Number 0-87891-579-6

Revised Printing 1989

WHAT "THE ESSENTIALS" WILL DO FOR YOU

This book is a review and study guide. It is comprehensive and it is concise.

It helps in preparing for exams, in doing homework, and remains a handy reference source at all times.

It condenses the vast amount of detail characteristic of the subject matter and summarizes the **essentials** of the field.

It will thus save hours of study and preparation time.

The book provides quick access to the important facts, principles, theorems, concepts, and equations of the field.

Materials needed for exams, can be reviewed in summary form – eliminating the need to read and re-read many pages of textbook and class notes. The summaries will even tend to bring detail to mind that had been previously read or noted.

This "ESSENTIALS" book has been carefully prepared by educators and professionals and was subsequently reviewed by another group of editors to assure accuracy and maximum usefulness.

Dr. Max Fogiel
Program Director

CONTENTS

This book covers the usual course outline for Calculus III. Earlier topics are covered in *"THE ESSENTIALS OF CALCULUS I"* and *"THE ESSENTIALS OF CALCULUS II"*.

Chapter No. **Page No.**

1 VECTOR ANALYSIS 1

1.1 Two Dimensional Vectors 1
1.1.1 Vector Properties 2
1.1.2 Addition of Two Vectors 2
1.1.3 Multiplication of a Vector by a Scalar 3
1.1.4 Additive and Multiplicative Properties of Vectors 3
1.1.5 Scalar (DOT) Product 4
1.2 Three Dimensional Vectors 4
1.2.1 Vector Properties 5
1.2.2 Linear Dependence and Independence 5
1.3 Vector Multiplication 7
1.3.1 Scalar (DOT) Product 7
1.3.2 Vector (CROSS) Product 8
1.3.3 Product of Three Vectors 11
1.4 Limits and Continuity 11
1.4.1 Limit of a Vector Function 12
1.5 Differentiation (Velocity, Acceleration and Arc Length) 12
1.6 Curvatures, Tangential and Normal Components 14
1.7 Kepler's Laws 15
1.7.1 First Law 15
1.7.2 Second Law 16
1.7.3 Third Law 16

2 REAL VALUED FUNCTIONS 17

2.1 Open and Closed Sets 17
2.2 Limits and Continuity 18

2.2.1 Definition of Continuity 19
2.3 Graphing 19
2.4 Quadric Surfaces 20

3 PARTIAL DIFFERENTIATION 23

3.1 Limits and Continuity 23
3.2 Partial Derivatives 23
3.2.1 Notation of Partial Derivatives 24
3.2.2 Higher-order Partial Derivatives 24
3.3 Increments and Differentials 25
3.4 Application of the Chain Rule 26
3.5 Directional Derivative and Gradients 27
3.5.1 Directional Derivative 27
3.5.2 Gradient 28
3.6 Tangent Planes 29
3.7 Total Differential 30
3.8 Taylor's Theorem with Remainder 31
3.8.1 Taylor's Theorem 31
3.9 Maxima and Minima 32
3.10 Lagrange Multipliers 34
3.11 Exact Differentials 36

4 MULTIPLE INTEGRATION 38

4.1 Double Integrals: Iterated Integrals 38
4.1.1 Properties of the Double Integral 39
4.1.2 Iterated Integrals 40
4.2 Area and Volume 42
4.2.1 Volume 42
4.2.2 Area 42
4.2.3 Volume of Surface of Revolution 42
4.3 Moment of Inertia and Center of Mass 43
4.3.1 Moments of Inertia 44
4.4 Polar Coordinates 44
4.5 The Triple Integrals 45
4.5.1 Application of Triple Integrals 46
4.6 Cylindrical and Spherical Coordinates of Triple Integrals 48
4.6.1 Cylindrical Coordinates 48
4.6.2 Spherical Coordinates 48
4.7 Surface Area A 49
4.8 Improper Integrals 49

5 **VECTOR FIELDS** 51

5.1 Vector Fields 51
5.2 Line Integrals 51
5.3 Independence of Path 52
5.4 Green's Theorem 53
5.5 Divergence and Curl 54

6 **INFINITE SERIES** 55

6.1 Indeterminate Forms 55
6.1.1 The Mean Value Theorem 55
6.1.2 L'Hôpital's Rule ($\frac{0}{0}$) 55
6.1.3 L'Hôpital's Rule ($\frac{\infty}{\infty}$) 56
6.2 Infinite Sequence 56
6.2.1 Properties 57
6.3 Convergent and Divergent Series 58
6.4 Positive Term Series 59
6.4.1 The Integral Test 60
6.4.2 The P-Series 60
6.4.3 Comparison Test 60
6.4.4 The Limit Comparison Test 60
6.5 Alternating Series, Absolute and Conditional Convergence 61
6.5.1 Alternating Series Test 61
6.5.2 Absolute Convergence 62
6.5.3 Conditional Convergence 62
6.5.4 Ratio Test 62
6.5.5 Root Test 62
6.6 Power Series 63
6.6.1 Calculus of Power Series 63
6.6.2 The Differentiation of Power Series 63
6.6.3 Integrating Term by Term 64
6.7 Taylor Series 65
6.7.1 Validity of Taylor's Expansion and Computations with Series 65
6.7.2 Binomial Theorem 66

CHAPTER 1

VECTOR ANALYSIS

1.1 TWO DIMENSIONAL VECTORS

Definition 1:

A scalar is a quantity that can be specified by a real number. It has only magnitude.

Definition 2:

A vector is a quantity that has both magnitude and direction. Velocity is an example of a vector quantity.

A vector (AB) is denoted by $\vec{AB}$, where B represents the head and A represents the tail. This is illustrated in Fig. 1.1.

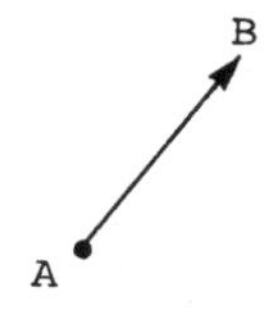

Fig. 1.1

The length of a line segment is the magnitude of a vector.

If the magnitude and direction of two vectors are the same, then they are equal.

Definition 3:

Vectors that can be translated from one position to another without any change in their magnitude or direction are called free vectors.

Definition 4:

The unit vector is a vector with a length (magnitude) of one.

Definition 5:

The zero vector has a magnitude of zero.

Definition 6:

The unit vector $\vec{i}$ is a vector with magnitude of one in the direction of the x-axis.

Definition 7:

The unit vector $\vec{j}$ is a vector with magnitude of one in the direction of the y-axis.

1.1.1 VECTOR PROPERTIES

1) When two vectors are added together, the resultant force of the two vectors produces the same effect as the two combined forces. This is illustrated in Fig. 1.2

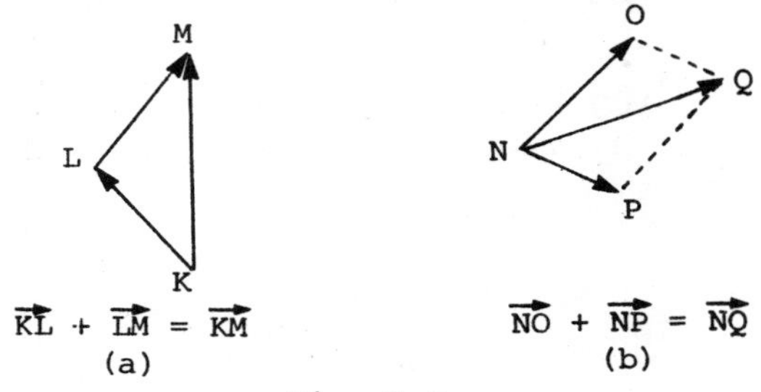

Fig. 1.2

In these diagrams, the vectors $\vec{KM}$ and $\vec{NQ}$ are the resultant forces.

1.1.2 ADDITION OF TWO VECTORS

Let vector A be $\langle a_1, a_2 \rangle$ and vector B be $\langle b_1, b_2 \rangle$. Then

$$A + B = (a_1+b_1)\vec{i} + (a_2+b_2)\vec{j}$$

1.1.3 MULTIPLICATION OF VECTOR BY A SCALAR

Let vector A be $a\vec{i} + b\vec{j}$ and let c be a constant. Then,

$$cA = c(a\vec{i}+b\vec{j}) = ca\vec{i} + cb\vec{j}$$

1.1.4 ADDITIVE AND MULTIPLICATIVE PROPERTIES OF VECTORS

Let s, t and u represent vectors and d and c represent real constants. All of the following are true:

1. $s + t = t + s$
2. $(s+t) + u = s + (t+u)$
3. $s + 0 = s$
4. $s + (-s) = 0$
5. $(c+d)s = cs + sd$
6. $c(s+u) = cs + cu$
7. $c(st) = (cs)t$
8. $1 \cdot s = s$
9. $0 \cdot s = 0$
10. $s \cdot s = |s|^2$
11. $c(ds) = (cd)s$

The magnitude $|s|$ of a vector $\vec{s} = a_1 i + a_2 j$ is

$$|s| = \sqrt{a_1^2 + a_2^2}$$

The difference between vectors $\vec{a}$ and $\vec{b}$ is given by the formula

$$\vec{a} - \vec{b} = \vec{a} + (-\vec{b})$$

1.1.5 SCALAR (DOT) PRODUCT

Two vectors are parallel if (a) one is a scalar multiple of the other; and (b) neither is zero.

Definition:

If vector A $= \langle a_1, a_2 \rangle$ and vector B $= \langle b_1, b_2 \rangle$, then the scalar product of A and B is given by the formula

$$A \cdot B = a_1 b_1 + a_2 b_2$$

Theorem:

If θ is the angle between the vectors $A = a_1 \vec{i} + a_2 \vec{j}$ and $B = b_1 \vec{i} + b_2 \vec{j}$ then

$$\cos\theta = \frac{a_1 b_1 + a_2 b_2}{|A||B|}$$

Definition:

Let vector $A = a_1 \vec{i} + a_2 \vec{j}$ and vector $B = b_1 \vec{i} + b_2 \vec{j}$ The projection of vector A on B ($\text{Proj}_B A$) is given by the quantity $|A| \cos\theta$, where θ is the angle between the two vectors.

Therefore,
$$\text{Proj}_B A = |A| \cos\theta = \frac{a_1 b_1 + a_2 b_2}{|B|} = \frac{A \cdot B}{|B|}$$

If the angle θ is acute, then $|A| \cos\theta$ is positive; if θ is obtuse, then $|A| \cos\theta$ is negative.

The scalar product of two non-zero vectors A and B is now redefined by the formula

$$A \cdot B = |A| |B| \cos\theta = a_1 b_1 + a_2 b_2$$

1.2 THREE DIMENSIONAL VECTORS

A vector in three-dimensional space is denoted by

$$\vec{v} = ai + bj + ck$$

where $\vec{i}$ is the unit vector in the direction of the x-axis, $\vec{j}$ is the unit vector in the direction of the y-axis, and $\vec{k}$ is the unit vector in the direction of the z-axis.

The magnitude of vector $\vec{v}$ is given by the formula

$$\boxed{|v| = \sqrt{a^2 + b^2 + c^2}}$$

The unit vector $\vec{u}$ in the direction of vector $\vec{v}$ is represented by

$$\boxed{\vec{u} = \frac{1}{|v|}\vec{v}}$$

1.2.1 VECTOR PROPERTIES

1) Two vectors are proportional (parallel) if each vector is a scalar multiple of the other (there is a number c such that $\vec{u} = c\vec{v}$).

2) Two vectors A and B are orthogonal (perpendicular) if and only if $A \cdot B = 0$.

3) Two vectors $\vec{AB}$ and $\vec{CD}$ are equal (have the same magnitude and direction) if and only if one of the following conditions is satisfied:

   a) The two vectors are on the same directed line $\vec{L}$ and their directed lengths are equal; or

   b) The points A, B, C, D are the vertices of a parallelogram. This is illustrated in Fig. 1.3.

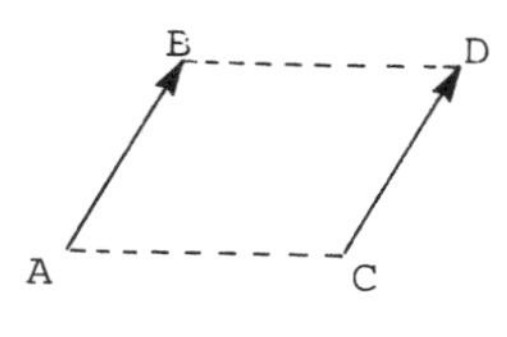

Fig. 1.3

1.2.2 LINEAR DEPENDENCE AND INDEPENDENCE

Let $\vec{v}_1, \vec{v}_2, \ldots, \vec{v}_n$ represent a set of vectors and $c_1, c_2, \ldots, c_n$ represent numbers.

An expression of the form

$$\boxed{c_1\vec{v}_1 + c_2\vec{v}_2 + \ldots + c_n\vec{v}_n}$$

is the linear combination of the vectors.

Two proportional vectors have a linear combination equal to the zero vector.

A set of vectors is linearly dependent if and only if there exists a set of constants such that

$$\boxed{c_1\vec{v}_1 + c_2\vec{v}_2 + \ldots + c_n\vec{v}_n = 0}$$

If these constants are all zero, then the set of vectors is said to be linearly independent. Two proportional (parallel) vectors are linearly dependent; thus one member of the set can be expressed as a linear combination of the remaining members.

A set of vectors, r,s,t, are linearly independent if

$$r = a_{11}i + a_{12}j + a_{13}k,$$

$$s = a_{21}i + a_{22}j + a_{23}k,$$

$$t = a_{31}i + a_{32}j + a_{33}k,$$

and the determinant $D = \begin{vmatrix} a_{11} & a_{12} & a_{13} \\ a_{21} & a_{22} & a_{23} \\ a_{31} & a_{32} & a_{33} \end{vmatrix}$

is not equal to zero.

The determinant of three vectors is found by expanding the vectors as follows:

$$\det A = \sum_{j=1}^{n} (-1)^{1+j} a_{ij} \det M_{ij}.$$

or, for the three vectors listed above,

$$\det\begin{bmatrix} a_{11} & a_{12} & a_{13} \\ a_{21} & a_{22} & a_{23} \\ a_{31} & a_{32} & a_{33} \end{bmatrix} = a_{11}\det M_{11} - a_{12}\det M_{12} + a_{13}\det M_{13}.$$

for an illustrated example, take the vectors

$$\vec{u} = 2i + j - k,$$
$$\vec{v} = -i + j + 2k, \quad \text{and}$$
$$\vec{w} = 2i + j + 3k.$$

$$D = \begin{vmatrix} 2 & 1 & -1 \\ -1 & 1 & 2 \\ 2 & 1 & 3 \end{vmatrix} = 2\begin{vmatrix} 1 & 2 \\ 1 & 3 \end{vmatrix} - 1\begin{vmatrix} -1 & 2 \\ 2 & 3 \end{vmatrix} - 1\begin{vmatrix} -1 & 1 \\ 2 & 1 \end{vmatrix}$$

$$= 2(3-2) - (-3-4) - (-1-2)$$

$$= 2 + 7 + 3 = 12$$

$D = 12 \neq 0$ thus the set of vectors is linearly independent.

Remember: The determinant of a 2×2 matrix $\begin{vmatrix} a & b \\ c & d \end{vmatrix}$ is found by $D = ad - bc$.

1.3 VECTOR MULTIPLICATION

1.3.1 SCALAR (DOT) PRODUCT

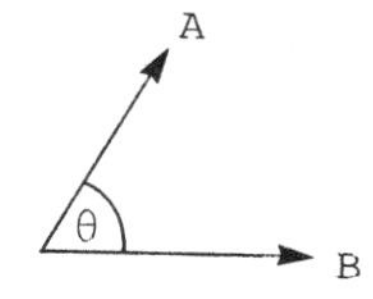

Let $A = a_1 i + a_2 j + a_3 k$ and $B = b_1 i + b_2 j + b_3 k$. If θ is the angle between these two vectors, then

$$\cos\theta = \frac{a_1 b_1 + a_2 b_2 + a_3 b_3}{|A|\,|B|}$$

The scalar product of vectors A and B is

$$\mathbf{A} \cdot \mathbf{B} = |\mathbf{A}|\ |\mathbf{B}| \cos\theta = a_1 b_1 + a_2 b_2 + a_3 b_3$$

Definition:

Let θ represent the angle between vectors A and B.

The component of A along B, or the projection of A on B ($|\mathbf{A}|\cos\theta$), is given by the formula

$$\text{Proj}_{\mathbf{B}} \mathbf{A} = |\mathbf{A}|\cos\theta = |\mathbf{A}| \frac{\mathbf{A} \cdot \mathbf{B}}{|\mathbf{A}|\ |\mathbf{B}|} = \frac{\mathbf{A} \cdot \mathbf{B}}{|\mathbf{B}|}$$

Scalar products are used to calculate the work done by a constant force when its point of application moves along a segment from C to D. The work done is the product of the distance from C to D and the projection of the constant force F on vector $\overrightarrow{CD}$.

$$W = \text{Proj}_{\overrightarrow{CD}} \mathbf{F} = \frac{\mathbf{F} \cdot \overrightarrow{CD}}{|CD|}$$

The work done by F is given by the formula

$$W = \mathbf{F} \cdot \mathbf{s}$$

1.3.2 VECTOR (CROSS) PRODUCT

The linearly independent vectors illustrated in Fig. 1.4 are said to form a right-handed triple. The vectors in Fig. 1.5 form a left-handed triple.

Fig. 1.4 Fig. 1.5

If two sets of ordered triples of vectors are both right-handed or left-handed, then they are said to be similarly oriented. If they are not, they are said to be oppositely oriented.

Theorem:

If the ordered triple <A,B,C> is right-handed, then the ordered triples <A,B, C> and <c_1A,c_2B,c_3C> are also right-handed, provided that $c_1, c_2, c_3 > 0$.

Definition:

If A and B are vectors, then the vector product A × B is defined as follows:

1) If either A or B is 0, then

$$\boxed{A \times B = 0}$$

2) If A is parallel to B, then

$$\boxed{A \times B = 0}$$

3) Otherwise,

$$\boxed{A \times B = C}$$

where vector C has the following properties:

a) It is orthogonal to both A and B.

b) It has magnitude $|C| = |A||B|\sin\theta$, where θ is the angle between A and B.

c) It is directed so that <A,B,C> is a right-handed triple.

Theorem:

Let

1) A and B represent any vector;

2) <i,j,k> represent a right-handed triple; and

3) t represent any number.

Then:

1) $A \times B = -(B \times A)$

2) $(tA) \times B = t(A \times B) = A \times (tB)$

3) $i \times j = -j \times i = k$

4) $j \times k = -k \times j = i$

5) $k \times i = -i \times k = j$

6) $i \times i = j \times j = k \times k = 0$

Theorem:

If A, B and D are any vectors, then:

1) $A \times (B+D) = A \times B + A \times D$

2) $(A+D) \times B = A \times B + A \times D$

Theorem:

If

$A = a_1 i + a_2 j + a_3 k$ and $B = b_1 i + b_2 j + b_3 k$

then the vector cross product $A \times B$ is given by

$$A \times B = (a_2 b_3 - a_3 b_2)i + (a_3 b_1 - a_1 b_3)j + (a_1 b_2 - a_2 b_1)k,$$

$$A \times B = \begin{vmatrix} i & j & k \\ a_1 & a_2 & a_3 \\ b_1 & b_2 & b_3 \end{vmatrix}$$

Example:

Find the cross product $A \times B$, if

$A = -2i + 4j + 5k$ and $B = 4i + 5k$.

$$A \times B = \begin{matrix} i & j & k & i & j & k \\ -2 & 4 & 5 & -2 & 4 & 5 \\ 4 & 0 & 5 & 4 & 0 & 5 \end{matrix}$$

$$\begin{array}{r} 20i + 20j + 0k \\ 0i + 10j - 16k \\ \hline \end{array}$$

$A \times B = 20i + 30j - 16k$

When moving upwards, multiply by -1.

1.3.3 PRODUCT OF THREE VECTORS

Theorem:

If $\vec{v}_1, \vec{v}_2, \vec{v}_3$ are vectors and the points P,Q,R,S are chosen so that

$$\vec{u}(\vec{PQ}) = \vec{v}_1, \quad \vec{u}(\vec{PR}) = \vec{v}_2 \text{ and } \vec{u}(\vec{PS}) = \vec{v}_3,$$

then:

1) $|(\vec{v}_1 \times \vec{v}_2)\vec{v}_3|$ is the volume of the parallelepiped with a vertex at P and adjacent vertices at Q, R and S. The volume is zero if and only if the four points P, Q, R and S lie in the same plane.

2) If $\langle i,j,k \rangle$ is a right-handed coordinate triple and if

$$\vec{v}_1 = a_1 i + b_1 j + c_1 k,$$

$$\vec{v}_2 = a_2 i + b_2 j + c_2 k,$$

$$\vec{v}_3 = a_3 i + b_3 j + c_3 k,$$

then

$$(\vec{v}_1 \times \vec{v}_2) \cdot \vec{v}_3 = \begin{vmatrix} a_1 i & b_1 j & c_1 k \\ a_2 & b_2 & c_2 \\ a_3 & b_3 & c_3 \end{vmatrix}$$

3) $(\vec{v}_1 \times \vec{v}_2) \cdot \vec{v}_3 = \vec{v}_1 \cdot (\vec{v}_2 \times \vec{v}_3)$.

Theorem:

Suppose A, B, and D are any vectors. Then:

1) $(A \times B) \times D = (A \cdot D) \cdot B - (B \cdot D) \cdot A$

2) $A \times (B \times D) = (A \cdot D) \cdot B - (A \cdot B) \cdot D$

1.4 LIMITS AND CONTINUITY

A vector function is a function that associates a unique element of a set of vectors with each element of a set of real numbers. The vector function is represented by v or $\vec{v}$.

If v is a vector function, then for each real number n in the domain of v, there exists a unique vector $v(t) \leq \langle x, y, z \rangle$.

A curve in three-dimensional space is a set of ordered triples of the form $\langle f(n), g(n), h(n) \rangle$ where the functions f, g and h are continuous on the interval I.

1.4.1 LIMIT OF A VECTOR FUNCTION

Definition:

If $V(n) = \langle f(n), g(n), h(n) \rangle$, then

$$\lim_{n \to a} v(n) = \langle \lim_{n \to a} f(n), \lim_{n \to a} g(n), \lim_{n \to a} h(n) \rangle$$

provided that f, g and h have limits as n approaches a.

Definition:

A vector function v is continuous at a if

$$\lim_{n \to a} v(n) = v(a).$$

1.5 DIFFERENTIATION (VELOCITY, ACCELERATION AND ARC LENGTH)

Definition:

If v is a vector function then we define the derivative v' as

$$v'(t) = \lim_{h \to 0} \frac{v(t+h)-v(t)}{h}$$

whenever the limit exists. If v'(t) exists, then we say that the function v is differentiable at t.

We may also denote the derivative as

$$\boxed{v'(t) = D_t v(t) = \frac{d}{dt}\, v(t)}$$

Theorem:

Let v and w represent differentiable vector functions and let c represent a scalar.

1) $D_t[v(t)+w(t)] = v'(t) + w'(t)$

2) $D_t[cv(t)] = cv'(t)$

3) $D_t[v(t) \cdot w(t)] = v(t) \cdot w'(t) + w(t) \cdot v'(t)$

4) $D_t[v(t) \times w(t)] = v(t) \times w'(t) + v'(t) \times w(t)$

5) $D_t[\,|v(t)|\,] = \dfrac{1}{|v(t)|}\,[v(t) \cdot v'(t)]$

(In all cases, $v(t) \neq 0$.)

The velocity vector $v(t_0)$ is a tangent vector to a curve in a plane.

The curve is represented parametrically as $x = x(t)$ at time $t = t_0$.

At any instant of time the velocity vector points in the direction of motion.

Thus,

$$v(t_c) = x'(t_0) = \lim_{t \to t_0} \frac{x(t) - x(t_0)}{t - t_0}$$

The speed at $t = t_0$ (the magnitude of the velocity) is represented by the formula

$$\boxed{|v(t_0)| = \sqrt{x'_1(t_0)^2 + x'_2(t_0)^2 + \ldots + x^1_n(t_0)}^{\,2}}$$

Definition:

The acceleration of a particle at time t_0 is

$$a(t_0) = \lim_{t \to t_0} \frac{v(t)-v(t_0)}{t-t_0} = v'(t_0) = x''(t_0)$$

Definition:

The arc length of a curve between t = a and t = b is

$$L = \int_a^b |x'(t)| \, dt$$

Speed measures the rate of change in arc length with respect to time.

1.6 CURVATURES, TANGENTIAL AND NORMAL COMPONENTS

The unit tangent vector to a curve x(t) is defined as

$$T(t) = \frac{1}{|x'(t)|} x'(t) \quad \text{if } x'(t) \neq 0$$

Definition:

The curvature k of a curve f(t) is defined as the magnitude of the rate of change of the direction of the curve with respect to arc length.

$$k = \left| \frac{dT(t)}{ds} \right|_{t = t_0} = \left| \frac{T'(t)}{f'(t)} \right|$$

where T(t) is the limit tangent vector.

f(t) is the equation of the curve given parametrically.

If $T' \neq 0$, then the unit normal vector to the curve is given by the formula

$$N = \frac{1}{|T'|} T'$$

The radius of a curvature is $R(t) = \frac{1}{k}$. The osculating plane is the plane k which contains the tangent line to the path and the center of curvature at t. This quantity is defined by only those values of t for which $k \neq 0$.

Thus,

$$N_1 = \frac{1}{k}\frac{dT}{ds}$$

The acceleration vector can be resolved into two components, the tangent to the curve and the normal to the curve.

$$a_T = \frac{d^2s}{dt^2} T$$

This formula expresses the acceleration in terms of its tangential component.

$$a_N = \left(\frac{ds}{dt}\right)^2 kN$$

This is the expression for the normal component of the acceleration.

Thus, the formula that expresses the acceleration in terms of its tangential and normal component is

$$a = \frac{d^2s}{dt^2} T + \left(\frac{ds}{dt}\right)^2 kN$$

1.7 KEPLER'S LAWS

Kepler's laws were formulated to describe the motion of planets about the sun.

1.7.1 FIRST LAW

The orbit of each planet is an ellipse with the sun at one focus.

1.7.2 SECOND LAW

The vector from the Sun to a moving planet sweeps over a fixed area at a constant rate.

1.7.3 THIRD LAW

If the time required for a planet to travel once around its elliptical orbit is T, and if the major axis of ellipse is 2a, then $T_2 = ka^3$ for some constant T.

Kepler's laws can be proven by the use of vector techniques since the force of gravity which the sun exerts on a planet exceeds that exerted by other celestial bodies. The orbit of a planet is a plane curve.

CHAPTER 2

REAL VALUED FUNCTIONS

2.1 OPENED AND CLOSED SETS

Boundary points: A point (a,b) in two-dimensional space is a boundary point of region R if every possible circle with center (a,b) contains both points which are in R and points which are not in R.

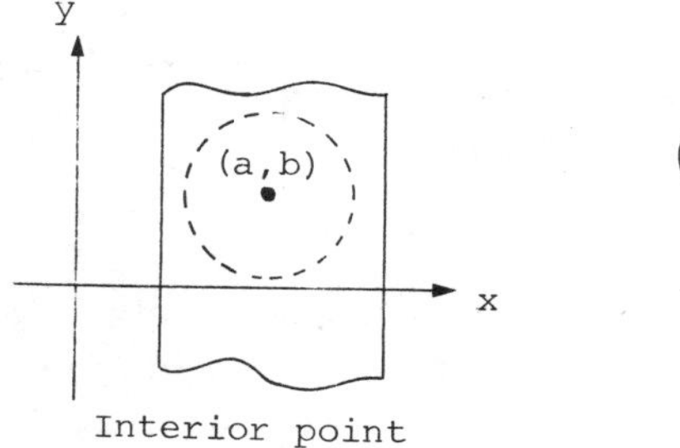

Interior point

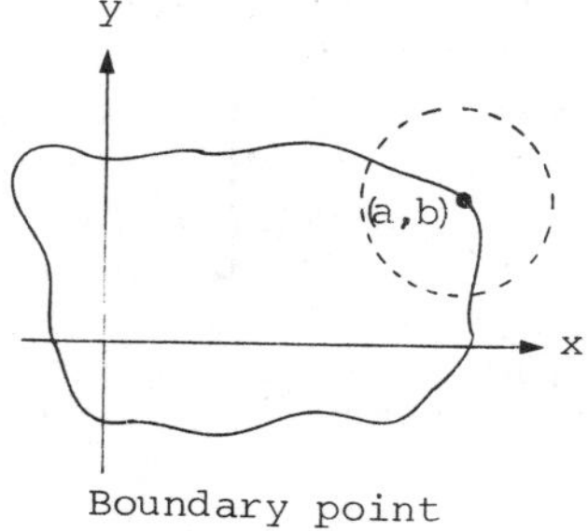

Boundary point

Interior Points: A point is an interior point of region R in two-dimensional space. If there exists a circle which has the point (a,b) as its center and which contains only points in the region R.

Open set: An open set is a set in which none of its points are boundary points.

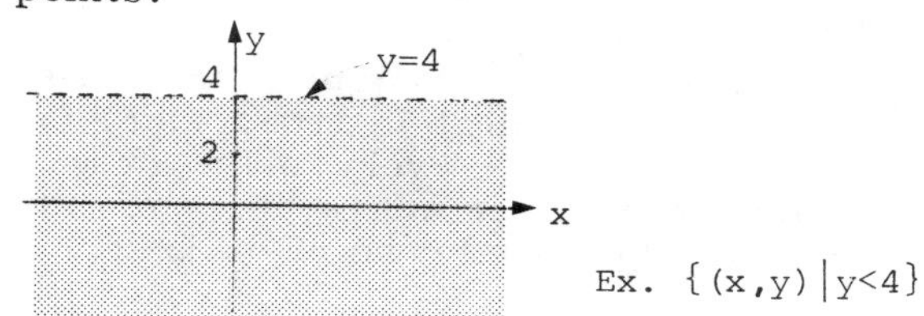

Closed set: A closed set includes all of its boundary points.

Ex. $\{(x,y)\,|\,x^2+y^2 \leq 4\}$

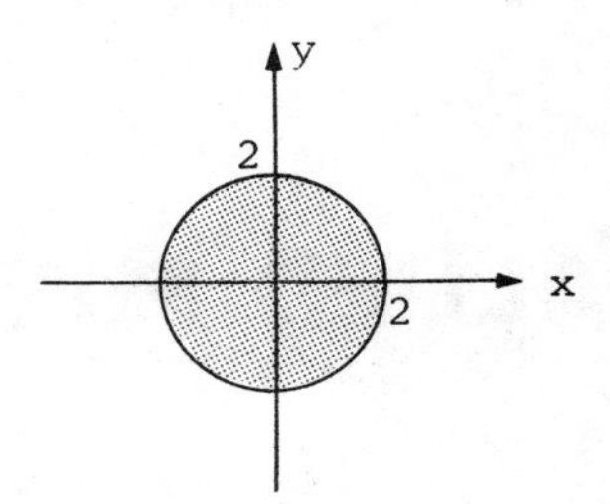

2.2 LIMITS AND CONTINUITY

The limit of f(x,y) as (x,y) approaches (a,b) is L. This is written:

$$\lim_{(x,y)\to(a,b)} f(x,y) = L$$

It means that for every $\varepsilon > 0$ there corresponds a $\delta > 0$ such that if $0 < \sqrt{(x-a)^2+(y-b)^2} < \delta$, then $|f(x,y)-L| < \varepsilon$.

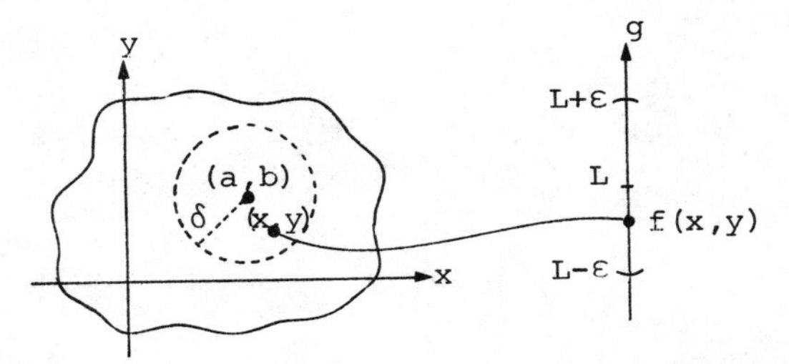

The preceding figures illustrate that, given that the definition of a limit is true, there is a circle of radius $\delta > 0$, such that for every point (x,y) inside the circle with radius δ and center (a,b), the number corresponding to f(x,y) is in the interval $(L-\varepsilon,\ L+\varepsilon)$.

If the limiting values obtained by taking two different paths to a point M(a,b) are different, then the limit of the function as (x,y) approaches (a,b) does not exist.

2.2.1 DEFINITION OF CONTINUITY

$$\lim_{(x,y)\to(a,b)} f(x,y) = f(a,b)$$

for f continuous at (a,b)

If a function f of two variables is continuous at (a,b) and a function g of one variable is continuous at f(a,b), then the function h, defined by $h(x,y) = g(f(x,y))$ is continuous at (a,b).

2.3 GRAPHING

Definition:

A cylinder is the set of all points on all lines which intersect a curve C in a plane and are parallel to a line L that is not in the plane.

Definition:

A surface of revolution is the surface which results from the revolution of a plane curve about a line in the plane.

Example: When a circle is revolved about a line along a diameter of the circle a sphere results.

If a parabola is revolved about its principal axis, the resulting surface is a paraboloid.

If a hyperbola is revolved about its transverse axis, the resulting surface is a hyperboloid of one sheet. If it is revolved about its conjugate axis, a hyperboloid of two sheets results.

When an ellipse is revolved about its major or minor axis the resulting surface is an ellipsoid.

Level Curves: Graphical representation of curves of the form $f(x,y) = k$, where k is a constant.

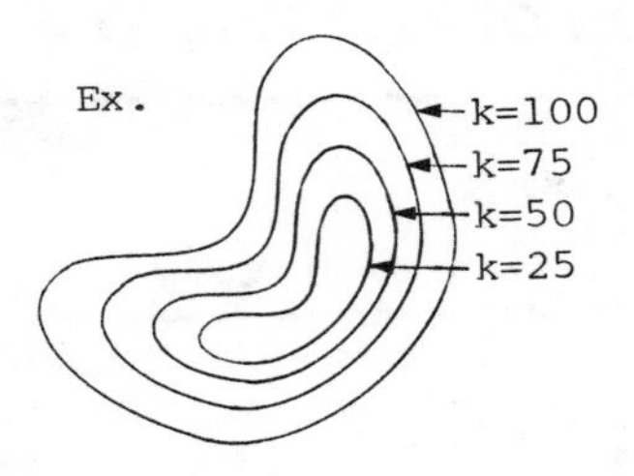

Level Surfaces: Graphical representation of curves of the form $f(x,y,z) = c$, where c is a constant.

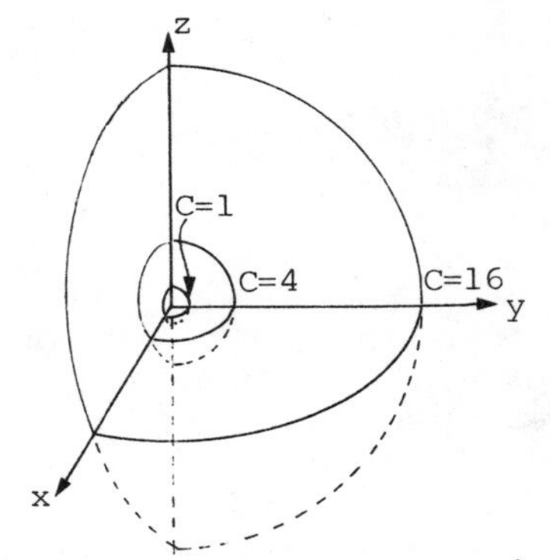

For the function $f(x,y,z,)=x^2+y^2+z^2=$ a constant C. As C becomes larger, we can picture an enlarging sphere.

2.4 QUADRIC SURFACES

Definition:

In three-dimensional analytic geometry, a quadric

surface is the graph of a second-degree equation in x, y, and z.

The graph of

$$\frac{x^2}{a^2} + \frac{y^2}{b^2} + \frac{z^2}{c^2} = 1$$

is an ellipsoid where a, b, and c are positive real numbers.

The graph of

$$\frac{x^2}{a^2} + \frac{y^2}{b^2} - \frac{z^2}{c^2} = 1$$

is a hyperboloid of one sheet.

The graph of

$$\frac{x^2}{a^2} - \frac{y^2}{b^2} - \frac{z^2}{c^2} = 1$$

is a hyperboloid of two sheets.

The graph of

$$\frac{x^2}{a^2} + \frac{y^2}{b^2} - \frac{z^2}{c^2} = 0$$

is a cone with z-axis as its axis.

The graph of

$$\frac{x^2}{a^2} + \frac{y^2}{b^2} = cz$$

is a paraboloid with the z-axis as its axis.

If $c > 0$, the paraboloid opens upward.

If $c < 0$, the paraboloid opens downward.

Graphs of equations of the form

$$\frac{x^2}{a^2} + \frac{z^2}{b^2} = cy \quad \text{and} \quad \frac{y^2}{a^2} + \frac{z^2}{b^2} = cx$$

are paraboloids whose axes are the y- and x-axis, respectively.

The graph of

$$\frac{y^2}{a^2} - \frac{x^2}{b^2} = cz$$

is a hyperbolic paraboloid.

Diagrams for the graphs of the above given equations can be found in any standard text covering the relevant material.

(1)

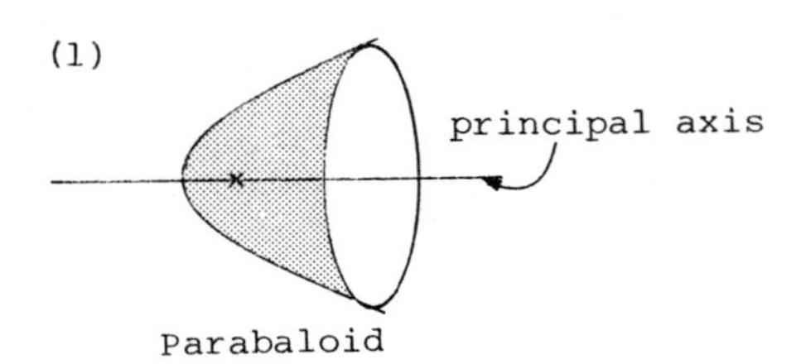

Parabaloid

(2)

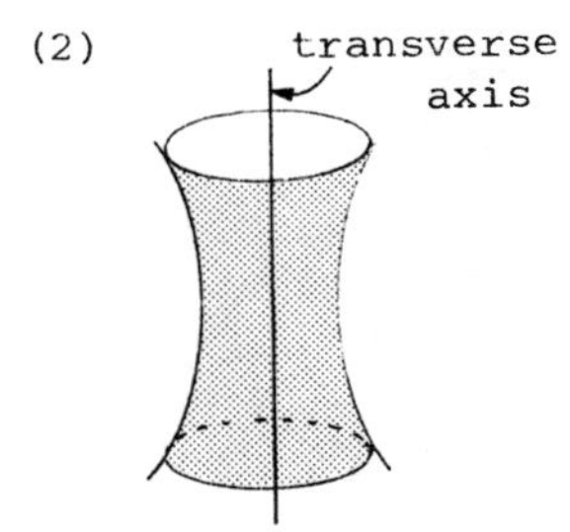

Hyperboloid of one sheet

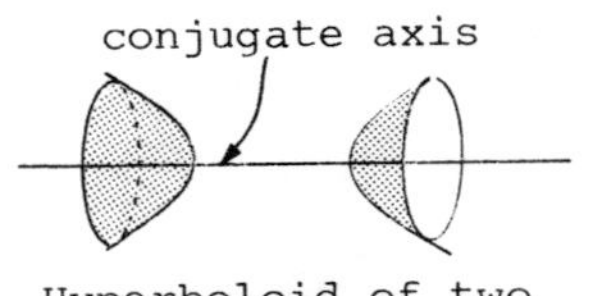

Hyperboloid of two sheets

(3)

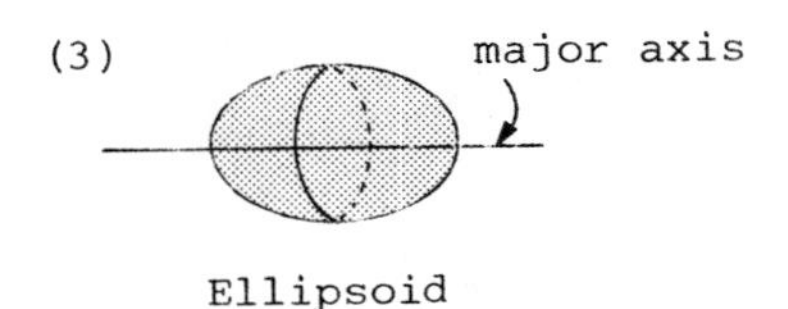

Ellipsoid

CHAPTER 3

PARTIAL DIFFERENTIATION

3.1 LIMITS AND CONTINUITY

A function is a mapping which takes each element of the domain into one and only one element of the range.

A function is a function of two variables if the domain consists of ordered pairs of numbers. A function of two variables is a polynomial function if $f(x,y)$ can be expressed as a sum of terms $cx^a y^b$, where c is a real number and a and b are non-negative integers.

Definition:

Let $f: x \to R$ represent a function of n variables and let a represent a point in the domain x of f. Then f is continuous at point a if

$$\lim_{x \to a} f(x) = f(a).$$

3.2 PARTIAL DERIVATIVES

Definition:

Let f represent a function of two variables. The first partial derivative of f with respect to x and y are the functions f_x and f_y, respectively. These functions are defined as

$$f_x(x,y) = \lim_{h\to 0} f\frac{(x+h,y) - f(x,y)}{h}$$

and

$$f_y(x,y) = \lim_{h\to 0} f\frac{(x,y+h) - f(x,y)}{h}$$

This holds true provided that the limits exist.

The two first partial derivatives are found in the manner prescribed below:

a) The derivative $f_x(x,y)$ is found by differentiating $f(x,y)$ with respect to x in the usual manner, while y is considered to be constant.

b) The derivative $f_y(x,y)$ is found by differentiating $f(x,y)$ with respect to y, keeping x constant.

3.2.1 NOTATION OF PARTIAL DERIVATIVES

1. $f_x = \frac{\partial f}{\partial x} = D_x f$ 2. $f_y = \frac{\partial f}{\partial y} = D_y f$

The derivative $f_x(x,y)$ is the measure of the rate of change of the function $f(x,y)$ as (x,y) moves in the horizontal direction.

The derivative $f_y(x,y)$ is the measure of the rate of change of the function $f(x,y)$ as (x,y) moves in the vertical direction.

The first partial derivative of functions of three or more variables is defined in the same manner except that all variables, except one, are held constant and differentiation takes place with respect to the remaining variable.

3.2.2 HIGHER-ORDER PARTIAL DERIVATIVES

The notations for the second partial derivatives are as follows:

1) $\frac{\partial}{\partial x} f_x = (f_x)_x = f_{xx} = \frac{\partial}{\partial x}\left(\frac{\partial f}{\partial x}\right) = \frac{\partial^2 f}{\partial x^2}$

2) $\frac{\partial}{\partial y} f_x = (f_x)_y = f_{xy} = \frac{\partial}{\partial y}\left(\frac{\partial f}{\partial x}\right) = \frac{\partial^2 f}{\partial y \partial x}$

3) $\frac{\partial}{\partial x} f_y = (f_y)_x = f_{yx} = \frac{\partial}{\partial x}\left(\frac{\partial f}{\partial y}\right) = \frac{\partial^2 f}{\partial x \partial y}$

4) $\frac{\partial}{\partial y}(f_y) = (f_y)_y = f_{yy} = \frac{\partial}{\partial y}\left(\frac{\partial f}{\partial y}\right) = \frac{\partial^2 f}{\partial y^2}$

The notations for the third partial derivatives are as follows:

1. f_{xxx} 2. f_{xxy} 3. f_{xyy} 4. f_{yyy}

5. f_{yyx} 6. f_{yxx} 7. f_{xyx} 8. f_{yxy}

The symbol f_{xyy} means that the order of the partial derivative is taken from left to right. The derivative with respect to x is taken first, then the derivative with respect to y is taken twice.

3.3 INCREMENTS AND DIFFERENTIALS

The increment of x and y is denoted by Δx and Δy, respectively, if f is a function of the two variables x and y.

Definition:

Notations of the increments:

$$f_x(x,y) = \lim_{\Delta x \to 0} \frac{f(x+\Delta x, y) - f(x,y)}{\Delta x}$$

$$f_y(x,y) = \lim_{\Delta y \to 0} \frac{f(x, y+\Delta y) - f(x,y)}{\Delta y}$$

If $z = f(x,y)$, then the differentials dx and dy of the independent variables x and y are defined as

$$dx = \Delta x \text{ and } dy = \Delta y.$$

The differential dz of the dependent variable z is defined as

$$dz = f_x(x,y)dx + f_y(x,y)dy$$

$$= \frac{\partial z}{\partial x} dx + \frac{\partial z}{\partial y} dy$$

If the first partial derivatives f_x and f_y are continuous on a rectangular region, then the function f is differentiable and therefore continuous.

Let us consider w, x and y independent variables and z, the dependent variable.

The differential of z is defined as

$$\boxed{dz = \frac{\partial z}{\partial x} dx + \frac{\partial z}{\partial y} dy + \frac{\partial z}{\partial w} dw}$$

3.4 APPLICATION OF THE CHAIN RULE

Partial derivatives are computed by using the chain rule.

Let $z = f(x,y)$ represent a continuous function and let $\frac{\partial f}{\partial x}$ and $\frac{\partial f}{\partial y}$ also be continuous.

If $x = x(a,b)$ and $y = y(a,b)$ are functions of a and b so that $\frac{\partial x}{\partial a}$, $\frac{\partial x}{\partial b}$, $\frac{\partial y}{\partial a}$, $\frac{\partial y}{\partial b}$ all exist, then z is a function of a and b and the following are true:

$$\frac{\partial z}{\partial a} = \left(\frac{\partial f}{\partial x}\right)\left(\frac{\partial x}{\partial a}\right) + \left(\frac{\partial f}{\partial y}\right)\left(\frac{\partial y}{\partial a}\right)$$

$$\frac{\partial z}{\partial b} = \left(\frac{\partial f}{\partial y}\right)\left(\frac{\partial y}{\partial b}\right) + \left(\frac{\partial f}{\partial x}\right)\left(\frac{\partial x}{\partial b}\right)$$

In the chain rule, a and b are considered independent variables; x and y are the intermediate variables.

The chain rule may be remembered as the rule by which the derivative appears in a fractional form.

This is illustrated by

$$\frac{dy}{dx} = \frac{dy}{dr} \cdot \frac{dr}{dx}$$

3.5 DIRECTIONAL DERIVATIVE AND GRADIENTS

3.5.1 DIRECTIONAL DERIVATIVE

Let f(x,y) represent a function and P(x,y) represent a point in the xy plane. A particular direction is found by specifying the angle θ, that the line through the point makes with the x-axis.

Definition:

If f is a function of x and y and $\vec{a} = \cos\theta i + \sin\theta j$ is the unit vector, then the directional derivative of f in the direction of a, denoted D_{af}, is given by the formula

$$D_{af} = \lim_{h \to 0} \frac{f(x+h\cos\theta, y+h\sin\theta)-f(x,y)}{h}$$

whenever the limit exists.

When $\theta = 0$, then $\cos\theta = 1$, $\sin\theta = 0$ and the direction is the positive x-direction. The directional derivative is $\partial f/\partial x$.

If θ is equal to $\pi/2$, we have $\cos\theta = 0$, $\sin\theta = 1$, and the directional derivative is $\frac{\partial f}{\partial y}$.

Theorem:

If $f(x,y)$ and its partial derivatives are continuous, and the unit vector a is equal to $\cos\theta i + \sin\theta j$, then $D_{af}(x,y) = f_x(x,y)\cos\theta + f_y(x,y)\sin\theta$.

An alternate notation for directional derivatives of functions of two variables is given by the formula

$$d\theta f(x,y)$$

where θ is the angle the direction makes with the positive x-axis.

The directional derivative is the function θ if the variables x and y have fixed values.

3.5.2 GRADIENT

The gradient of a function is a vector containing the partial derivatives of the function.

Definition:

If $f(x,y)$ has partial derivatives, then the gradient is defined as

$$\text{grad } f(x,y) = f_x(x,y)i + f_y(x,y)j.$$

The symbol "del" (∇) is used to denote a gradient.

The gradient $f(x,y)$ may also be denoted by $\nabla f(x,y)$, where

$$\nabla = i\frac{\partial}{\partial x} + j\frac{\partial}{\partial y}$$

The dot product of a gradient and a unit vector is illustrated below:

$$\vec{a} \cdot \nabla f = |a|\,|\nabla f| \cos\phi = D_{af}$$

where a is the unit vector,

∇f is the gradient

ϕ is the angle between the vector $\vec{a}$ and the gradient ∇f.

We may now conclude that the directional derivative is maximum when ϕ is zero, that is when the unit vector is in the direction of the gradient.

3.6 TANGENT PLANES

Let P represent a point on the graph of the equation f(x,y,z). The first partial derivatives of this function are continuous and are not all zero at point P.

The derivative at point P is the slope of the line tangent to the curve at this point. The plane through point P, with normal vector ∇f, is the tangent plane at that point.

Definition:

The equation of the tangent plane to the graph of $f(x,y,z) = 0$ at the point $P(x_0,y_0,z_0)$ is

$$f_x(x_0,y_0,z_0)(x-x_0)+f_y(x_0,y_0,z_0)(y-y_0)+f_z(x_0,y_0,z_0)(z-z_0) = 0$$

The normal line is the line perpendicular to the tangent plane at point P. The equation of the normal line is

$$\frac{x - x_0}{f_x(x_0,y_0,z_0)} = \frac{y - y_0}{f_y(x_0,y_0,z_0)} = \frac{z - z_0}{f_z(x_0,y_0,z_0)}$$

We conclude that the normal line is perpendicular to the tangent plane.

3.7 TOTAL DIFFERENTIAL

Given the function $y = f(x)$, the quantity df is the differential of f defined as

$$df = f'(x)h,$$

where h and x are independent variables.

Definition:

If f is a function of three variables x,y,z, then the total differential is defined as

$$dF(x,y,z) = F_x(x,y,z)h + F_y(x,y,z)k + F_z(x,y,z)\ell$$

Theorem:

Suppose that $z = f(x,y)$ and x and y are functions of the same variable. Then,

$$dz = \frac{\partial z}{\partial x} dx + \frac{\partial z}{\partial y} dy.$$

This is similar to the results obtained for the function $w = f(x,y,z)$.

$$dw = \frac{\partial w}{\partial x} dx + \frac{\partial w}{\partial y} dy + \frac{\partial w}{\partial z} dz$$

This equation holds when x, y and z are either independent or intermediate variables.

Example: Find the derivative dy/dx of the following equation by methods of partial differentiation:

$$e^{xy} + \sin xy + 1 = 0$$

$$f_x = ye^{xy} + y \cos xy$$

$$f_y = xe^{xy} + x \cos xy$$

$$\frac{dy}{dx} = -\frac{ye^{xy} + y \cos xy}{xe^{xy} + x \cos xy} = -\frac{y(e^{xy} + \cos xy)}{x(e^{xy} + \cos xy)}$$

$$\frac{dy}{dx} = \frac{-y}{x}$$

3.8 TAYLOR'S THEOREM WITH REMAINDER

The expansion of the function f(x) with n+1 derivatives on an interval containing the value x_0 is

$$F(x) = F(x_0) + F'(x_0)(x-x_0)+\ldots+ \frac{F^{(n)}(x_0)(x-x_0)^n}{n!} + Rn$$

Rn is the remainder represented by the formula

$$Rn = \frac{F^{(n+1)}(\varepsilon_0)(x-x_0)^{n+1}}{(n+1)!}$$

where ε_0 is some number between x and x_0.

3.8.1 TAYLOR'S THEOREM

Let f be a continuous function of two variables and also let all of its partial derivatives be continuous about the point P(a,b).

The expansion for f is

$$F(x,y) = F(a,b) = \sum_{1 \leq r+s \leq p} \frac{\partial^{r+s} f(a,b)}{\partial x^r \partial y^s} \frac{(x-a)^r}{r!} \cdot \frac{(y-b)^s}{s!} + Rp.$$

The remainder R_p is represented by the formula

$$R_p = \sum_{r+s \pm p+1} \frac{\partial^{r+s} F(\varepsilon,\eta)}{\partial_x^r \partial_y^s} \frac{(x-a)^r}{r!} \frac{(y-b)^s}{s!}$$

The value of (ε,η) is located on the line segment between points (a,b) and (x,y).

The remainder term demonstrates that if $0 < r < d$, then

$$\varepsilon = a + \frac{(x-a)}{d}\tau, \quad \eta = b + \frac{(y-b)}{d}\tau .$$

Another mode of expression of Taylor's rule is

$$F(x,y) = f(a,b) + \sum_{g=1}^{p} \frac{1}{q!} \left[\sum_{r=0}^{q} \frac{q!}{(g-r)!r!} \frac{\partial^q f(a,b)}{\partial x^{q-r} \partial y^r} (x-a)^{q-r} (y-b)^r \right] + Rp$$

If P tends to infinity for some function f, the remainder approaches 0 ($R_p \to 0$) and we obtain a representation of f as an infinite series in x and y. This series is called a double series and f is said to have expanded about the point (a,b).

3.9 MAXIMA AND MINIMA

A function f(x,y) of two variables is said to have a relative (local) maximum at (x_0,y_0) if there is a rectangular region (R) containing (x,y) such that $F(x,y) \leq f(x_0,y_0)$ for all pairs (x,y) in the rectangular region.

The relative maximum can be attained geometrically by drawing the graph of the function. The high points on the graph represent the relative maxima.

Theorem:

Let f(x,y) represent a function in the rectangular region R. Let $f_x(x_0,y_0)$ and $f_y(x_0,y_0)$ be defined, and let $f(x,y) \leq f(x_0,y_0)$ for all (x,y) in R.

If

$$f_x(x_0,y_0) = f_y(x_0,y_0) = 0,$$

then $f(x_0,y_0)$ is the relative maximum.

The function f(x,y) has a relative minimum at (a,b) if there is a region containing (a,b) such that $f(x,y) \geq f(a,b)$ for all (x,y) in the region. The relative minimum corresponds to the low point on the graph of the function.

Let f(x,y) represent a continuous function on the closed rectangular region. The function has an absolute maximum f(a,b) and an absolute minimum f(c,d) for some (a,b) and (c,d) in the region. Thus,

$$f(c,d) \leq f(x,y) \leq f(a,b)$$

for all (x,y) in the rectangular region.

Definition:

The point on a graph at which both f_x and f_y vanish is called a critical point.

A critical point at which the function f is neither a maximum nor a minimum is a saddle point. This is illustrated in Fig. 3.1.

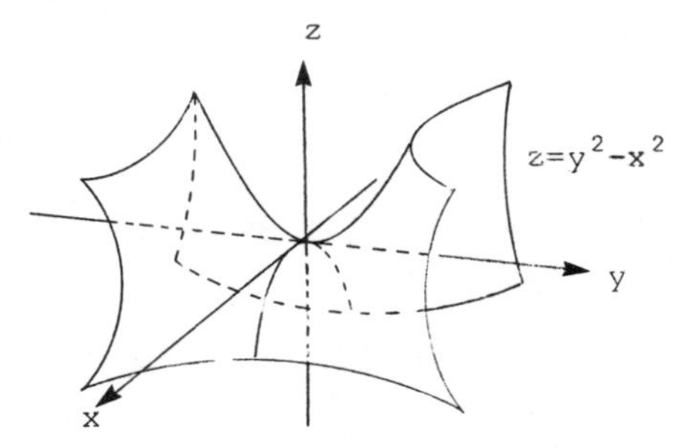

Fig. 3.1 In this graph, the saddle point is (0,0,0)

Theorem: Test for Extrema

Let f(x,y) and its partial derivatives to the third order be continuous near the point (a,b).

The point (a,b) is a critical point if $f_x(a,b) = f_y(a,b) = 0$. Thus,

1) f(a,b) is a local minimum if $f_{xx}(a,b)f_{yy}(a,b) - [f_{xy}(a,b)]^2 > 0$ and $f_{xx}(a,b) > 0$.
2) f(a,b) is a local maximum if $f_{xx}(a,b)f_{yy}(a,b) - [f_{xy}(a,b)]^2 > 0$ and $f_{xx}(a,b) < 0$.

3) $f(a,b)$ is a saddle point if $f_{xx}(a,b)f_{yy}(a,b) - [f_{xy}(a,b)]^2 < 0$.

4) No information is attained if $f_{xx}(a,b)f_{yy}(a,b) - [f_{xy}(a,b)]^2 = 0$.

If the function $f(x,y,z)$ has first derivatives, then a critical point exists at the point where $f_x = f_y = f_z = 0$.

3.10 LAGRANGE MULTIPLIERS

Problems of free maxima and minima are problems in which the maxima and minima are functions of several variables without any added conditions.

A problem involving an added or side condition is called a problem in constrained maxima and minima.

Let f and g be functions of x, y and z. To find the local extrema of $f(x,y,z)$ subject to the constraint $\phi(x,y,z) = 0$:

1) We solve the equation $\phi(x,y,z)$ to obtain $z = g(x,y)$, where the functions f, ϕ and g have continuous first partial derivatives throughout a suitable domain.

2) We define a function of four variables by $w = f(x,y,z) + h\phi(x,y,z)$, where h is the Lagrange multiplier.

3) The values of x,y,z which give the extrema of f are obtained by solving four equations in four unknowns, simultaneously.

For example, if $w = u + h\nu$, then the extremas are given by the solution of the following four equations in four unknowns:

$$w_x = u_x + h\nu_x = 0,$$

$$w_y = u_y + h\nu_y = 0,$$

$$w_z = u_z + h\nu_z = 0,$$

and $w_h = \nu = 0$.

We assume that the partial derivatives of ν are not zero, and solve the first three equations for h to obtain

$$h = \frac{-u_x}{\nu_x} = \frac{-u_y}{\nu_y} = \frac{-u_z}{\nu_z}$$

Example: Find the minimum of $f(x,y,z) = x^2 + y^2 + z^2$, subject to the condition $x + 3y - 2z - 4 = 0$.

$$w(x,y,z,h) = x^2 + y^2 + z^2 + h(x+3y-2z-4) = 0 \quad (1)$$

$$w_x = 2x + h = 0 \quad (2)$$

$$w_y = 2y + 3h = 0 \quad (3)$$

$$w_z = 2z - 2h = 0 \quad (4)$$

From equations (2), (3) and (4) we obtain

$$h = -2x = -\frac{2}{3}y = z,$$

which gives

$$x = \frac{1}{3}y \quad \text{and} \quad z = -\frac{2}{3}y.$$

Substitute these into

$$x + 3y - 2z - 4 = 0,$$

and we have

$$\frac{1}{3}y + 3y - 2\left(-\frac{2}{3}y\right) - 4 = 0,$$

$$y = 2.$$

Thus, $\quad x = \frac{1}{3}y = \frac{1}{3}\times 2 = \frac{2}{3}$ and

$$z = -\frac{2}{3}y = -\frac{2}{3}\times 2 = -\frac{4}{3}.$$

The minimum of $f(x,y,z)$ is then found by:

$$f(x,y,z) = x^2 + y^2 + z^2$$

$$= \left(\frac{2}{3}\right)^2 + 2^2 + \left(-\frac{4}{3}\right)^2$$

$$= \frac{4}{9} + 4 + \frac{16}{9} = \frac{56}{9}.$$

3.11 EXACT DIFFERENTIALS

The differential of the function f(x,y) is

$$df = \frac{\partial f}{\partial x} dx + \frac{\partial f}{\partial y} dy$$

The quantity df is a function of four variables.

Definition:

Suppose there is a function f(x,y,z) such that

$$df = P(x,y,z)dx + Q(x,y,z)\,dy + R(x,y,z)dz$$

for all (x,y,z) in some region, and for all values of dx, dy, dz, we say that

$$Pdx + Qdy + Rdz$$

is the exact differential.

Example: Show that $(2xe^{x^2}\sin y)dx + (e^{x^2}\cos y)dy$ is an exact differential and find the function f of which it is the total differential.

$$(2xe^{x^2}\sin y)dx + (e^{x^2}\cos y)dy$$

Solution

Let $P = 2xe^{x^2}\sin y$, and $Q = e^{x^2}\cos y$. Differentiate to find P_y and Q_x:

$$P_y = 2xe^{x^2}\cos y, \quad Q_x = 2xe^{x^2}\cos y$$

$$P_y = Q_x = 2xe^{x^2} \cos y,$$

Thus,the function is an exact differential.

Next, set $f_x = 2xe^{x^2} \sin y$. Thus,

$$f = e^{x^2} \sin y + c(y).$$

Differentiate with respect to y to obtain

$$f_y = e^{x^2} \cos y + c'(y)$$

Let f_y equal to Q:

$$e^{x^2} \cos y + c'(y) = e^{x^2} \cos y$$

Solve for $c'(y)$:

$$c'(y) = 0$$

Integrate to obtain $c(y)$:

$$c(y) = c = \text{constant}$$

Thus,

$$f(x,y) = e^{x^2} \sin y + c.$$

CHAPTER 4

MULTIPLE INTEGRATION

4.1 DOUBLE INTEGRALS: ITERATED INTEGRALS

Definition:

Suppose f is a function defined on a region R. The double integral of f over R, denoted by $\iint_R f(x,y)dA$, is given by the formula

$$\iint_R f(x,y)dA = \lim_{||P|| \to 0} \Sigma f(u_i, v_i) \Delta A_i ,$$

provided that the limit exists.

The double integral measures the volume under a surface region.

Theorem:

If the function f(x,y) is continuous for (x,y) in a closed region R, then f is said to be over R. Furthermore, if $f(x,y) > 0$ for (x,y) in R, then

$$v = \iint_R f(x,y)dA,$$

where v is the volume of the solid defined by

$$B = \{(x,y,z):(x,y) \text{ in } R, \text{ and } 0 \le z \le f(x,y)\}.$$

4.1.1 PROPERTIES OF THE DOUBLE INTEGRAL

1) Suppose c is a number and f is integrable over a closed region R, then cf is integrable and

$$\iint_R cf(x,y)\,dA = c\iint_R f(x,y)\,dA$$

2) If f and g are two integrable functions over a closed region R, then

$$\iint_R [f(x,y)+g(x,y)]\,dA = \iint_R f(x,y)\,dA + \iint_R g(x,y)\,dA$$

3) Consider f is an integrable function on a closed region R, and $m \le f(x,y) \le M$ for all (x,y) in R. If A denotes the area of R, we have

$$mA \le \iint_R f(x,y)\,dA \le MA$$

4) If the function f and g are both integrable over the region R, and $f(x,y) \le g(x,y)$ for all (x,y) in R, then

$$\iint_R f(x,y)\,dA \le \iint_R g(x,y)\,dA$$

5) Suppose that the closed region R is decomposed into non-overlapping regions R_1 and R_2, and suppose that f

is continuous over R, then:

$$\iint_R f(x,y)dA = \iint_{R_1} f(x,y)dA + \iint_{R_2} f(x,y)dA$$

4.1.2 ITERATED INTEGRALS

The symbol $\int_c^d f(x,y)dy$ denotes the partial integration with respect to y.

For each x in the interval [a,b], there corresponds a unique value of this integral.

A function A is thus determined, where the value A(x) is given by the formula

$$A(x) = \int_c^d f(x,y)dy.$$

The function A is continuous on the interval [a,b], and its definite integral may be written

$$\int_a^b A(x)dx = \int_a^b \left[\int_c^d f(x,y)dy \right] dx$$

The expression on the right side of the equation is called an iterated (double) integral.

The terms "successive integrals" and "repeated integrals" may also be used.

$$\int_c^d \int_a^b f(x,y)\,dx\,dy = \int_c^d \left[\int_a^b f(x,y)\,dx \right] dy$$

$$\int_a^b \int_c^d f(x,y)\,dy\,dx = \int_a^b \left[\int_c^d f(x,y)\,dy \right] dx$$

Iterated integrals may also be defined over regions that have curved boundaries.

Theorem:

Suppose R is a closed region given by

$$R = \{(x,y) : a \le x \le b,\ p(x) \le y \le g(x)\},$$

where p and g are continuous and $p(x) \le g(x)$ for $a \le x \le b$. Also suppose that $f(x,y)$ is continuous on R. Then,

$$\iint_R f(x,y)\,dA = \int_a^b \int_{p(x)}^{q(x)} f(x,y)\,dy\,dx$$

The corresponding results hold if the closed region R has the representation

$$R = \{(x,y) : c \le y \le d,\ r(y) \le x \le s(y)\},$$

where $r(y) \le s(y)$ for $c \le y \le d$. In such a case,

$$\iint_R f(x,y)\,dA = \int_c^d \int_{r(y)}^{s(y)} f(x,y)\,dx\,dy$$

Both iterated integrals are equal to the double integral and are therefore equal to each other.

4.2 AREA AND VOLUME

4.2.1 VOLUME

The volume of a region R is represented by the double integral of a non-negative function. This volume is expressed in terms of the volume of the cylinders which have generators parallel to the z-axis and located between the surface and the region R in the xy plane. Thus,

$$\text{Volume (v)} = \iint_R dA$$

4.2.2 AREA

The area of a surface is found by using the iterated integral. The formula used to find the area of a region is

$$A = \iint_R dA = \int_a^b \int_{p(x)}^{q(x)} dydx = \lim_{||\Delta||} \sum_i \sum_j \Delta y_j \, \Delta x_i$$

4.2.3 VOLUME OF SURFACE OF REVOLUTION

Theorem:

If a plane figure B lies on one side of a line L in its plane, the volume of the surfaces, generated by revolving the figure around the line, is equal to the product of the area (A) and the length of the path described by the center of mass. Thus, if the figure is in the xy plane and L is the x-axis, then,

$$v = 2\pi\bar{y}A = \iint 2\pi y \, dA$$

4.3 MOMENT OF INERTIA AND CENTER OF MASS

Let L denote a lamina having the shape of the region R as shown in the figure below.

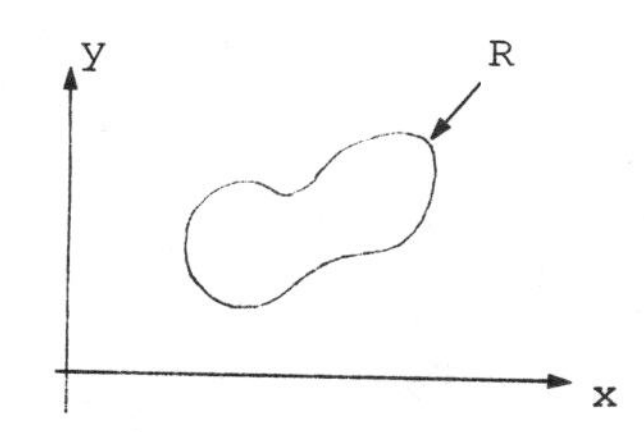

Let it be given that the density per unit area at a point (x,y) is $\rho(x,y)$ where ρ is a continuous function on R.

The mass of L is given by

$$M = \iint_R \rho(x,y)dA.$$

The moment of L with respect to the x-axis is given by

$$M_x = \iint_R y\ \rho(x,y)dA.$$

The moment of L with respect to the y-axis is given by

$$M_y = \iint_R x\,\rho(x,y)dA.$$

The center of mass of the Lamina, L, is the point $(\bar{x},\bar{y})$, where

$$\bar{x} = \frac{M_y}{M} \quad \text{and} \quad \bar{y} = \frac{M_x}{M}.$$

4.3.1 MOMENTS OF INERTIA

The moment of inertia with respect to the x-axis is

$$I_x = \sum_{i=1}^{n} y_i^2 m_i,$$

for n particles of masses $m_1, m_2, \ldots, m_n$ located at points $(x_1, y_1), (x_2, y_2), \ldots, (x_n, y_n)$, respectively.

Also the moment of inertia with respect to the y-axis is

$$I_y = \sum_{i=1}^{n} x_i^2 m_i.$$

For lamina:

$$I_x = \iint_R y^2 \rho(x,y)\,dA$$

and

$$I_y = \iint_R x^2 \rho(x,y)\,dA$$

4.4 POLAR COORDINATES

Points in the Cartesian coordinate system can be transformed to the polar coordinate system by use of the formulas

$$\boxed{x = r\cos\theta \quad \text{and} \quad y = r\sin\theta}$$

The area of a region in polar coordinates is given by the formula

$$A = \iint_R dA_{r,\theta} = \int_{\theta_1}^{\theta_2} \left[\int_{r_1}^{r_2} r dr \right] d\theta$$

where $dA_{r\theta} = r\, dr d\theta$

Theorem:

If a function $f(x,y)$ is continuous on a region R and if this region is related to another region R* by mapping $x = r\cos\theta$ and $y = r\sin\theta$, then the function $s(r,\theta) = f(r\cos\theta, r\sin\theta)$ is defined and continuous on R* and

$$\iint_R f(x,y) dA_{x,y} = \iint_{R^*} s(r,\theta)\ dA_{r,\theta}$$

This equation is transformed to

$$\iint_R f(x,y)\ dxdy = \iint_{R^*} s(r,\theta) r\ dr d\theta.$$

4.5 THE TRIPLE INTEGRALS

The triple integral of a function over the region R is represented by the formula

$$\iiint f(x,y,z) dv = \int_k^e \int_c^d \int_a^b f(x,y,z) dx, dy, dz.$$

The integral on the right is the iterated integral.

Triple integrals may be defined on regions bounded by a parallelepiped and regions in the xy plane.

Theorem:

Suppose that function f is a continuous function on the region R, which is defined by the inequalities

$$R = \{(x,y,z) : a \le x \le b,\ p(x) \le y \le q(x),$$
$$r(x,y) \le z \le s(x,y)\},$$

where the functions p, q, r and s are continuous. Then,

$$\iiint_R f(x,y,z)dv = \int_a^b \left\{ \int_{p(x)}^{q(x)} \left[\int_{r(x,y)}^{s(x,y)} f(x,y,z)dz \right] dy \right\} dx$$

4.5.1 APPLICATION OF TRIPLE INTEGRALS

The mass of a solid is represented by the formula

$$\boxed{m = \iiint_R \rho\,(x,y,z)dv \qquad \text{if:}}$$

1) The solid has the shape of a three-dimensional region R.
2) The density at (x,y,z) is $\rho(x,y,z)$ where ρ is continuous throughout the region.

The moment of a particle of mass at the point (x,y,z) is

$$M_{xy} = \iiint_R z\,\rho(x,y,z)dv,$$

$$M_{xz} = \iiint_R y\,\rho(x,y,z)dv,$$

and $$M_{yz} = \iiint_R x\,\rho(x,y,z)dv,$$

where the moment was taken with respect to the xy-, xz- and yz-planes.

The center of mass is represented by the point $(\bar{x},\bar{y},\bar{z},)$, where

$$\bar{x} = \frac{M_{yz}}{m}, \quad \bar{y} = \frac{M_{xz}}{m} \quad \text{and} \quad \bar{z} = \frac{M_{xy}}{m}.$$

The center of mass of a homogeneous solid is dependent only on the shape of the region, since the function ρ is constant and therefore is cancelled.

A centroid is defined as the graph of corresponding points for geometric solids in two dimensions.

The moments of inertia (I) of a mass with respect to the x-, y-, and z-axes are given by the following formulas:

$$I_x = \iiint_R (y^2+z^2)\,\rho(x,y,z)\,dv$$

$$I_y = \iiint_R (x^2+z^2)\,\rho(x,y,z)\,dv$$

$$I_z = \iiint_R (x^2+y^2)\,\rho(x,y,z)\,dv$$

The radius of gyration of a solid of mass m which has a moment of inertia (I) with respect to a line, is defined as a number r, such that

$$I = mr^2$$

Thus we conclude that the radius of gyration is the distance from the line at which all the mass could be concentrated without changing the moment of inertia of the solid.

4.6 CYLINDRICAL AND SPHERICAL COORDINATES OF TRIPLE INTEGRALS

4.6.1 CYLINDRICAL COORDINATES

Triple integrals are sometimes expressed in terms of cylindrical coordinates.

One such case occurs if a function of r, θ and z is continuous throughout a region

$$R = \{(r,\theta,z) : a \le r \le b,\ c \le \theta \le d,\ k \le z \le \ell\}.$$

If a point in the region R is represented by (r_i, θ_i, z_i), then the triple integral of the function over the region R is

$$\iiint_R f(r,\theta,z)dv = \lim_{\|P\|\to 0} \Sigma f(r_i,\theta_i,z_i)\Delta v_i,$$

where $||P||$ represents the length of the longest diagonal in the region.

$$\iiint_R f(r,\theta,z)dv = \int_k^\ell \int_c^d \int_a^b f(r,\theta,z)rdrd\theta\, dz$$

By using inner partitions, triple integrals may be defined over complicated regions.

Thus,

$$\iiint_R f(r,\theta,z)dv = \int_\alpha^\beta \int_{p(\theta)}^{q(\theta)} \int_{k_1(r,\theta)}^{k_2(r,\theta)} f(r,\theta,z)rdzdrd\theta$$

4.6.2 SPHERICAL COORDINATES

Triple integrals may also be expressed using spherical

coordinates. If a function f of ρ, ϕ and θ is continuous throughout a region

$$R = \{(\rho, \phi, \theta) = a \leq \rho \leq b,\ c \leq \theta \leq d,\ k \leq \phi \leq \ell\},$$

then

$$\iiint_R f(\rho, \phi, \theta)\,dv = \int_c^d \int_k^\ell \int_a^b f(\rho, \phi, \theta)\rho^2 \sin\phi \, d\rho \, d\phi \, d\theta .$$

The letter ρ does not denote density in the equation of spherical coordinates.

4.7 SURFACE AREA A

Let f represent a function which is defined as $f(x,y) \geq 0$ throughout a region R in the xy plane and which has continuous first derivatives on R.

T is defined as the part of the graph of f for which a projection on the xy-plane is R.

The area of T is defined by the formula

$$A = \iint_R \sqrt{(f_x(x,y))^2 + (f_y(x,y))^2 + 1} \; dA.$$

This formula is applicable when $f(x,y) \leq 0$ on the region R.

4.8 IMPROPER INTEGRALS

$$\int_a^\infty f(x)dx = \lim_{t \to \infty} \int_a^t f(x)dx,$$

if the limit exists

$$\int_{-\infty}^{a} f(x)dx = \lim_{t \to -\infty} \int_{t}^{a} f(x)dx,$$

if the limit exists.

The above two expressions are improper integrals since one of their limits is not a real number.

Improper integrals either converge or diverge

They converge when the limit on the right-hand side of the equation exists as t goes to infinity.

They diverge when the limit on the right-hand side of the equation does not exist as t goes to infinity.

$$\int_{-\infty}^{+\infty} f(x)dx = \int_{-\infty}^{a} f(x)dx + \int_{a}^{\infty} f(x)dx$$

The above integral converges if and only if both integrals on the right-hand side of the equation are convergent.

Another form of an improper integral is one in which the integrand does not exist for some value on the closed interval of integration.

Example: If f is continuous on the half-open interval $[a,b)$ and becomes infinite or undefined at b, then, by definition,

$$\int_{a}^{b} f(x)dx = \lim_{t \to b^-} \int_{a}^{t} f(x)dx,$$

provided the limit exists.

Likewise, if f is continuous on the half-open interval $(a,b]$ and becomes infinite or undefined at a, then by definition,

$$\int_{a}^{b} f(x)dx = \lim_{t \to a^+} \int_{t}^{b} f(x)dx,$$

provided the limit exists.

As previously stated, the above integral converges if the limit on the right-hand side of the expression exists, and diverges otherwise.

CHAPTER 5

VECTOR FIELDS

5.1 VECTOR FIELDS

Definition:

All the vectors in a region such that a unique vector having initial point P is associated with each point P in the region is called a vector field.

Examples of vector fields

1) Electric fields

2) Magnetic fields

A vector field that is the gradient of a scalar function is called a conservative vector field.

In the rectangular coordinate system the vector associated with the point $M(x,y,z)$ may be denoted by $F(x,y,z)$, where

$$F(x,y,z) = M(x,y,z)i + N(x,y,z)j + P(x,y,z)k.$$

Every equation of the type given above determines a vector field. The function $F(x,y,z)$ is called a vector function.

5.2 LINE INTEGRALS

Consider a plane curve given by the parametric equations

$$x = h(t) \text{ and } y = g(t) \quad (a \le t \le b),$$

where g and h are smooth on the interval [a,b].

Let it be given that the norm $||\Delta||$ of the subdivision of c is, by definition, the largest of the Δs_i, where Δs_i denotes the length of the subarc $\overline{P_{i-1}P_i}$.

Then the line integral along c from A → B is given by

$$\int_C f(x,y)ds = \lim_{||\Delta|| \to 0} \sum_i f(u_i v_i)\Delta s_i.$$

If f is continuous on the interval, then the above limit exists and we can rewrite the above as:

$$\int_C f(x,y)ds = \int_{t_1}^{t_2} f(h(t),g(t)) \sqrt{[h'(t)]^2+[g'(t)]^2}\, dt.$$

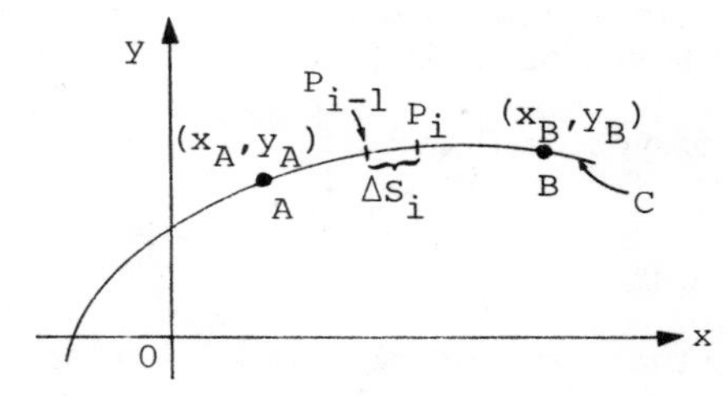

5.3 INDEPENDENCE OF PATH

Definition:

A path is a piecewise smooth curve connecting two points P and Q.

A line integral is independent of path if the same value of the line integral is obtained for all possible paths from P to Q.

The line integral $\int_C F \cdot dr$ is independent of path C if and only if $F(x,y) = \nabla f(x,y)$ (i.e., the vector field is conservative).

Example:

$$\int_{(x_1,y_1)}^{(x_2,y_2)} M(x,y)dx + N(x,y)dy = \int_{(x_1,y_1)}^{(x_2,y_2)} df = f(x_2,y_2) - f(x_1,y_1)$$

5.4 GREEN'S THEOREM

A smooth curve is one such that if represented parametrically by $x = g(t)$ and $y = h(t)$, g' and h' must be continuous on the closed interval $[a,b]$ and should not be zero simultaneously except possibly at a or b.

A smooth closed curve is one such that

$$A = (g(a),h(a)) = (g(b),h(b)) = B.$$

A simple curve is one such that for all numbers t_1 and t_2,

$$t_1 \neq t_2 \text{ and } (g(t_1),h(t_1)) \neq (g(t_2),h(t_2)).$$

GREEN'S THEOREM

Consider c, a piecewise smooth simple closed curve and R, the region consisting of c and its interior. If M and N are continuous function of x and y with continuous first partial derivatives in R, then

$$\oint_C Mdx + Ndy = \iint_R \left(\frac{\partial N}{\partial x} - \frac{\partial M}{\partial y} \right) dA.$$

5.5 DIVERGENCE AND CURL

In rectangular coordinates

$$\nabla = i \frac{\partial}{\partial x} + j \frac{\partial}{\partial y} + k \frac{\partial}{\partial z}$$

$$\text{grad } f = \nabla f = \frac{\partial f}{\partial x} i + \frac{\partial f}{\partial y} j + \frac{\partial f}{\partial z} k$$

The curl of F, denoted by curl F or $\nabla \times F$, is defined by

$$\text{curl } F = \nabla \times F = \begin{vmatrix} i & j & k \\ \frac{\partial}{\partial x} & \frac{\partial}{\partial y} & \frac{\partial}{\partial z} \\ M & N & P \end{vmatrix}$$

where $F(x,y,z) = M(x,y,z)i + N(x,y,z)j + P(x,y,z)k$.

The divergence of F, denoted by div F or $\nabla \cdot F$, is defined by

$$\text{div } F = \nabla \cdot F = \frac{\partial M}{\partial x} + \frac{\partial N}{\partial y} + \frac{\partial P}{\partial z}$$

The ratio of the closed-surface integral of a vector field taken about a small (vanishing) closed surface to the small volume enclosed by the closed surface is known as the divergence of the vector field.

The ratio of the closed-line integral of a field taken about a small (vanishing) closed path to the small area enclosed expressed as a vector is known as the curl of the field.

CHAPTER 6

INFINITE SERIES

6.1 INDETERMINATE FORMS

An indeterminate form is a ratio $\frac{f(x)}{g(x)}$ in which $f(c) = g(c) = 0$ or $f(c) = g(c) = \infty$ for some c. Even though $\frac{f(c)}{g(c)}$ is meaningless, $\lim \frac{f(x)}{g(x)}$ may exist in such cases. (See 6.1.2)

$$\frac{0}{0} \quad \text{and} \quad \frac{\infty}{\infty}.$$

The indeterminate form $\left(\frac{0}{0}\right)$

6.1.1 THE MEAN VALUE THEOREM

Suppose that f and g are differentiable functions on the interval [a,b] and that they are also continuous. If g' is never 0 in (a,b), then there is a number c in (a,b) such that

$$\frac{f'(c)}{g'(c)} = \frac{f(b)-f(a)}{g(b)-g(a)}$$

6.1.2 L'HOSPITAL'S RULE $(\frac{0}{0})$

Suppose that

$$\lim_{x\to c} f(x) = 0, \lim_{x\to c} g(x) = 0, \lim_{x\to c} \frac{f'(x)}{g'(x)} = \ell,$$

and that the Mean Value Theorem holds in some interval about c. Then,

$$\lim_{x \to c} \frac{f(x)}{g(x)} = \lim_{x \to c} \frac{f'(x)}{g'(x)} = \ell$$

6.1.3 L'HOSPITAL'S RULE $\left(\frac{\infty}{\infty}\right)$

Suppose that

$$\lim_{x \to c} f(x) = \infty, \quad \lim_{x \to c} g(x) = \infty, \quad \text{and} \lim_{x \to c} \frac{f'(x)}{g'(x)} = \ell$$

Then,

$$\lim_{x \to c} \frac{f(x)}{g(x)} = \lim_{x \to c} \frac{f'(x)}{g'(x)} = \ell$$

L'Hospital's rule is used for one-sided limits as well as for ordinary limits. You may have to apply L'Hospital's rule more than once.

6.2 INFINITE SEQUENCE

Definition:

An infinite sequence is a function whose domain is the set of positive integers.

Definition:

A sequence $\{a_n\}$ has the limit ℓ (denoted by $\lim_{n \to \infty} a_n = \ell$) if for every $c > 0$, there exists a positive number N such that if $n > N$, then $|a_n - \ell| < c$.

If the limit ($\lim_{n \to \infty} a_n$) does not exist, then the sequence

$\{a_n\}$ has no limit.

The statement $\lim_{n\to\infty} a_n = \infty$ means that for every positive real number s, there exists a number N such that if $n > N$, then $a_n > s$.

Theorem:

1) $\lim_{n\to\infty} r^n = 0$, if $|r| < 1$.

2) $\lim_{n\to\infty} |r^n| = \infty$, if $|r| > 1$.

6.2.1 PROPERTIES

Suppose that $\lim_{n\to\infty} a_n = \ell$ and $\lim_{n\to\infty} b_n = m$, where m and ℓ are real numbers, then it can be proven that:

1) $\lim_{n\to\infty} (a_n+b_n) = \ell + m$

2) $\lim_{n\to\infty} (a_n-b_n) = \ell - m$

3) $\lim_{n\to\infty} a_n b_n = \ell \cdot m$

4) $\lim_{n\to\infty} \frac{a_n}{b_n} = \frac{\ell}{m}$

Theorem:

If $\{a_n\}$, $\{b_n\}$ and $\{c_n\}$ are infinite sequences such that $a_n \le b_n \le c_n$ for all n, and if

$$\lim_{n\to\infty} a_n = \ell = \lim_{n\to\infty} c_n,$$

then $$\lim_{n\to\infty} b_n = \ell .$$

Theorem:

Suppose that $\{a_n\}$ represents a sequence and that $\lim_{n\to\infty} |a_n| = 0$ then $\lim_{n\to\infty} a_n = 0$.

A monotonic sequence is a sequence with successive terms that are non-decreasing

$$(a_1 \le a_2 \le \ldots \le a_n)$$

or non-increasing

$$(a_1 \le a_2 \ge \ldots \ge a_n).$$

A sequence is said to be bounded if there is a positive real number α such that $|a_m| \le \alpha$ for all m.

An infinite sequence that is bounded and monotonic is said to have a limit.

6.3 CONVERGENT AND DIVERGENT SERIES

Definition:

An infinite series $\sum_{k=0}^{\infty} a_k$, sometimes written $a_0+a_1+a_2 \ldots$, is the sequence $\{s_0, s_1, s_2, s_3, \ldots\}$ of partial sums.

An infinite series converges if $\lim_{n\to\infty} s_n = s$ for some real number s. The series diverges if the sequence of the partial sums diverges (the limit does not exist).

Geometric series have the form $a + ar + ar^2 + \ldots + ar^{n-1}$, where a and r are real numbers and $a \neq 0$.

Theorem:

The geometric series

$$a + ar + ar^2 + \ldots + ar^{n-1},$$

with $a \neq 0$:

1) converges and has the sum $\frac{a}{1-r}$ if $|r| < 1$.

2) diverges if $|r| \geq 1$.

If an infinite series Σa_n is convergent, then $\lim_{n\to\infty} a_n = 0$. The infinite series Σa_n is divergent if $\lim_{n\to\infty} a_n \neq 0$. If $\lim_{n\to\infty} a_n = 0$, this does not mean that the series is convergent.

Theorem:

For every $c > 0$, if there exists an integer N such that $|s_k - s_\ell| < c$ whenever $k, \ell > N$, then the infinite series Σa_n is convergent.

Theorem:

If Σa_n and Σb_n are infinite series such that $a_i = b_i$ for all $i > k$, where k is a positive integer, then both series converge or both series diverge.

Theorem:

If Σa_n and Σb_n are convergent series with the sums A and B, respectively, then:

1) $\Sigma(a_n + b_n)$ converges and has the sum $A + B$.

2) If c is a real number, Σca_n converges and has the sum cA.

3) $\Sigma(a_n - b_n)$ converges and has the sum $A - B$.

6.4 POSITIVE TERM SERIES

A positive term series is a series for which every term is positive.

Theorem:

If Σa_n is a positive term series and if there exists a number m such that $s_n < m$ for every n, then the series converges and has a sum $s \leq m$. If no such m exists, the series diverges.

6.4.1 THE INTEGRAL TEST

If a function f is positive, continuous and decreasing on the interval $[1,\infty)$, then the infinite series:

1) converges if $\int_1^\infty f(x)dx$ converges.

2) diverges if $\int_1^\infty f(x)dx$ diverges.

6.4.2 THE P-SERIES

The p-series expressed as $\sum_{n=1}^{\infty} \frac{1}{n^p}$, converges if $p > 1$ and diverges if $p \leq 1$.

6.4.3 COMPARISON TEST

Let Σa_n and Σb_n represent positive term series.

1) If $a_n \leq b_n$ and the series Σb_n converges, then Σa_n converges.
2) If $a_n \geq b_n$ and the series Σb_n diverges, then Σa_n also diverges.

6.4.4 THE LIMIT COMPARISON THEOREM

Let Σa_n and Σb_n represent positive term series.

If $\lim\limits_{n\to\infty} \frac{a_n}{b_n} = \ell$, where ℓ is some positive number, then either both series converge or both diverge.

6.5 ALTERNATING SERIES: ABSOLUTE AND CONDITIONAL CONVERGENCE

An alternating series is an infinite series in which successive terms have opposite signs. An alternating series is usually expressed as

$$a_1 - a_2 + a_3 - a_4 + \ldots + (-1)^{n-1} a_n + \ldots,$$

or

$$-a_1 + a_2 - a_3 + a_4 - \ldots + (-1)^n a_n + \ldots,$$

where each $a_i > 0$.

6.5.1 ALTERNATING SERIES TEST

Let $\{a_k\}$ represent a decreasing series of positive terms. If $a_k > 0$, then $\sum\limits_{k=0}^{\infty} (-1)^k a_k$ converges.

Theorem:

If the conditions $a_k > a_{k+1} > 0$ for every positive interger k hold for the alternating series $\Sigma(-1)^{k-1} a_k$ and $\lim\limits_{n\to\infty} a_k = 0$, then the error in approximating the sum S of the series by the k+n partial sum, S_k, is numerically less then a_{k+1}.

6.5.2 ABSOLUTE CONVERGENCE

The series Σa_k is absolutely convergent if the series $\Sigma |a_k| = |a_1| + |a_2| + \ldots + |a_n|$, obtained by taking the absolute value of each term, is convergent.

6.5.3 CONDITIONAL CONVERGENCE

The series Σa_k is conditionally convergent if $\Sigma |a_k|$ diverges while Σa_k converges.

6.5.4 RATIO TEST

Suppose that in the series Σa_k, every $a_k \neq 0$,

$$\lim_{k \to \infty} \left| \frac{a_{k+1}}{a_k} \right| = \rho \quad \text{or} \quad \lim_{k \to \infty} \left| \frac{a_{k+1}}{a_k} \right| = +\infty .$$

Then:

1) If $\rho < 1$, the series Σa_k converges absolutely.

2) If $\rho > 1$, or if $\lim\limits_{k \to \infty} \dfrac{|a_{k+1}|}{|a_k|} = +\infty$, the series diverges.

3) If $\rho = 1$, the test gives no information.

6.5.5 ROOT TEST

Let Σa_k represent an infinite series.

1) If $\lim\limits_{k \to \infty} \sqrt[k]{|a_k|} = \ell < 1$, the series is absolutely convergent.

2) If $\lim\limits_{k \to \infty} \sqrt[k]{|a_k|} = \ell > 1$ or $\lim\limits_{k \to \infty} \sqrt[k]{|a_k|} = \infty$, the series is divergent.

3) If $\lim\limits_{k \to \infty} \sqrt[k]{|a_k|} = 1$, the series may be absolutely convergent, conditionally convergent, or divergent.

6.6 POWER SERIES

A power series is a series of the form

$$c_0 + c_1(x-a)+c_2(x-a)^2 +\ldots+ c_n(x-a)^n$$

in which a and c_i, i = 1,2,3, etc. are constants.

The notations

$$\sum_{n=0}^{\infty} c_n(x-a)^n \quad \text{and} \quad \sum_{n=0}^{\infty} c_n x^n$$

are used to describe power series.

A power series $\Sigma c_n x^n$ is said to converge:

1) at x_1 if and only if $\Sigma c_n x^n$ converges.
2) on the set S if and only if $\Sigma c_n x^n$ converges for each $x \in S$.

If $\Sigma c_n x^n$ converges at $x_1 \neq 0$, then it converges absolutely whenever $|x| < |x_1|$. If $\Sigma c_n x^n$ diverges at x_1, then it diverges for $|x| > |x_1|$.

6.6.1 CALCULUS OF POWER SERIES

If the series $\Sigma c_n x^n$ converges on the interval (-a,a), then

$$\Sigma \frac{d}{dx}(c_n x^n) = \Sigma n c_n x^{n-1}$$

also converges on (-a,a).

6.6.2 THE DIFFERENTIATION OF POWER SERIES

If

$$f(x) = \sum_{n=0}^{\infty} c_n x^n \text{ for all x in } (-a,a),$$

then f is differentiable on (-a,a), and

$$f'(x) = \sum_{n=1}^{\infty} n c_n x^{n-1} \quad \text{for all } x \text{ in } (-a,a).$$

A power series defines an infinite differentiable function in the interior of its interval of convergence.

The derivatives of this function may be obtained by differentiating term by term.

6.6.3 INTEGRATING TERM BY TERM

If

$$f(x) = \Sigma c_n x^n$$

converges on the interval (-a,a), then

$$g(x) = \Sigma \frac{c_n}{n+1} x^{n+1}$$

converges on (-a,a), and

$$\int f(x)dx = g(x) + c.$$

The equations for term-by-term integration are expressed as either

$$\int (\Sigma c_n x^n)dx = \left(\Sigma \frac{c_n}{n+1} x^{n+1} \right) + c$$

for indefinite integrals or

$$\int_a^b (\Sigma c_n x^n)dx = \Sigma \left(\int_a^b c_n x^n dx \right)$$

$$= \Sigma \frac{c_n}{n+1} \left(b^{n+1} - a^{n+1} \right)$$

for definite integrals (provided [a,b] is contained in the interval of convergence).

6.7 TAYLOR SERIES

$$f(x) = \sum_{n=0}^{\infty} \frac{f^{(n)}(a)}{n!} (x-a)^n$$

This formula represents the Taylor series for f about the point a or the expansion of f into a power series about a.

For the special case where a = 0, the Taylor series is represented by the formula

$$f(x) = \sum_{n=0}^{\infty} \frac{f^{(n)}(0)}{n!} x^n$$

This is called the Maclaurin series.

6.7.1 VALIDITY OF TAYLOR'S EXPANSION AND COMPUTATIONS WITH SERIES

The Taylor series for e^x about x = a converges to e^x for any a and x. This is exemplified by the formula

$$e^x = e^a \sum_{n=0}^{\infty} \frac{(x-a)^n}{n!}$$

This equation is applicable for all values of a and x.

The following functions are obtained from the Maclaurin

series:

$$\sin x = \sum_{n=0}^{\infty} \frac{(-1)^n x^{2n+1}}{(2n+1)!},$$

$$\cos x = \sum_{n=0}^{\infty} \frac{(-1)^n x^{2n}}{(2n)!}$$

6.7.2 BINOMIAL THEOREM

For each real number m, we have

$$(1+x)^m = 1 + \sum_{n=1}^{\infty} \frac{m(m-1)(m-2)\ldots(m-n+1)}{n!} x^n \text{ for } |x| < 1.$$